SOLAR SYSTEM ACTIVITY BOOK

AUTHOR Linda Milliken
EDITOR Kathy Rogers
PAGE DESIGN Linda Milliken
ARTIST Barb Lorseyedi
COVER DESIGN Imaginings

METRIC CONVERSION CHART
Refer to this chart when metric conversions are not found within the activity.

¼ tsp	= 1.25 ml	350° F	=	175° C
½ tsp	= 2.5 ml	375° F	=	190° C
1 tsp	= 5 ml	400° F	=	200° C
1 Tbsp	= 15 ml	425° F	=	220° C
¼ cup	= 60 ml	1 inch	=	2.54 cm
⅓ cup	= 75 ml	1 foot	=	30 cm
½ cup	= 125 ml	1 yard	=	91 cm
1 cup	= 230 ml	1 mile	=	1.6 km
1 oz.	= 28 g			
1 lb.	= .45 kg			

www.edupressinc.com
ISBN 1-56472-124-8
Printed in USA

TABLE OF CONTENTS

GLOSSARY

astronomer—someone who studies the stars and planets and other objects in space.

atmosphere—the layer of gases that surrounds a planet or a star.

barred spiral—a kind of galaxy which has a bar of gas and stars running across its center and two arms closely circling the outside.

Big Bang—the huge explosion that many scientists think started the universe moving outward.

black hole—the remains of a star. It has very strong gravity; it sucks in every object around it, even rays of light, which is why we cannot see a black hole.

comet—small frozen mass of dust and gas revolving around the sun.

constellation—a group of stars as they are seen from the Earth.

dwarf—a very small star.

elliptical galaxy—type of galaxy which is shaped like an oval.

galaxy—group of millions and millions of stars which are all loosely held together by gravity.

giant—a very large star, much larger than our sun.

globular cluster—a group of very old stars that cling together in the shape of a ball.

gravity—the force that pulls objects toward each other.

gravity wave—a surge of force that may pass through space when a star explodes.

light-year—the distance light travels in one year, equal to about six trillion miles.

meteor—the streak of light observed when a meteoroid enters the Earth's atmosphere.

Milky Way—the edge of our galaxy as it appears from the Earth.

moon—a smaller body that travels around a planet.

nebula—a cloud of gas and dust in space.

orbit—a path through space followed by one thing around another.

phase—any stage in a series or cycle of changes, as of the moon's illumination.

planet—a body in space which moves around a star such as the sun.

quasar—an object far away in space that looks like a star but gives off immense quantities of light and radio waves.

red dwarf—a star that is one-tenth the size of the sun; red dwarfs are much cooler so they do not give off much light.

red giant—dying star that swells many times larger than its normal size.

rotate—to turn around an axis.

satellite—a small body in orbit around a larger body in space.

Seyfert galaxy—a kind of galaxy which gives off an unusual amount of light from its center.

solar—having to do with the sun.

solar system—the sun and all the objects that orbit it such as planets and moons.

spiral galaxy—a kind of galaxy with a flattened shape and arms curving out from the center.

star—a glowing ball of gas that gives off its own light and heat.

star cluster—a group of stars that are very close together.

supercluster—a huge group of many galaxies that are close together.

supernova—the explosion of a very big star at the end of its life.

universe—all of space and everything in it.

white dwarf—the remains of a dead star which is very small.

GLOSSARY GAME

Challenge your knowledge of solar system words. Draw a line from each word to its definition. Exchange papers with a classmate to check your answers.

orbit	the force that pulls objects toward each other
galaxy	an object far away in space that looks like a star but gives off immense quantities of light and radio waves
quasar	a group of stars as they are seen from the Earth
gravity	the remains of a star that has such strong gravity that it sucks in everything around it, including light rays
moon	a path through space followed by one thing around another
black hole	a group of millions and millions of stars which are all loosely held together by gravity
nebula	the streak of light that is observed when a meteoroid enters the Earth's atmosphere
meteor	a cloud of dust and gas in space
constellations	a smaller body that travels in an orbit around a planet

UNIVERSE

INFORMATION

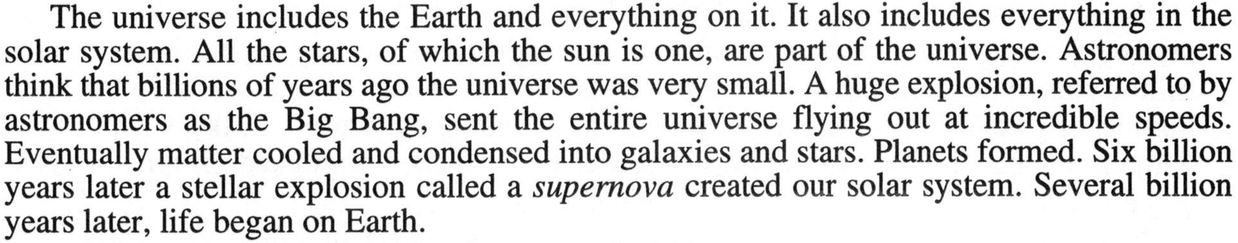

The universe includes the Earth and everything on it. It also includes everything in the solar system. All the stars, of which the sun is one, are part of the universe. Astronomers think that billions of years ago the universe was very small. A huge explosion, referred to by astronomers as the Big Bang, sent the entire universe flying out at incredible speeds. Eventually matter cooled and condensed into galaxies and stars. Planets formed. Six billion years later a stellar explosion called a *supernova* created our solar system. Several billion years later, life began on Earth.

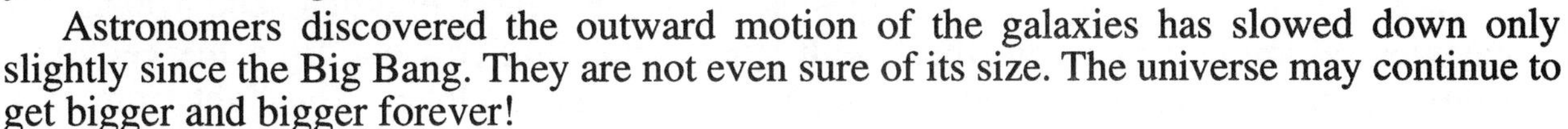

Astronomers discovered the outward motion of the galaxies has slowed down only slightly since the Big Bang. They are not even sure of its size. The universe may continue to get bigger and bigger forever!

PROJECT

Conduct experiments relating to the origin and future of the universe.

MATERIALS

Origin
- Air popcorn popper
- Popcorn

Future
- Balloons
- Marking pens

ORIGIN—BIG BANG

1. Place popcorn in the popper. Observe and discuss the position of the kernels. Leaving the lid off, turn on the popper. Maintain a safe distance.
2. Note sounds heard and changes observed in the kernels. Develop cause and effect theories. When the popping is complete, observe the position of the kernels. Describe the changes in size and position.
3. Relate the results of the experiment to the Big Bang theory.

FUTURE

1. Use a marking pen to draw small "stars" or dots on the balloon.
2. Blow up the balloon slightly. Observe the relationship of the dots. What is causing them to move apart? Continue to blow up the balloon a little at a time. Observe and draw scientific conclusions that relate to the universe today.
3. What eventually happens to the balloon? Can classroom astronomers offer any theories for the future?

GALAXY

INFORMATION

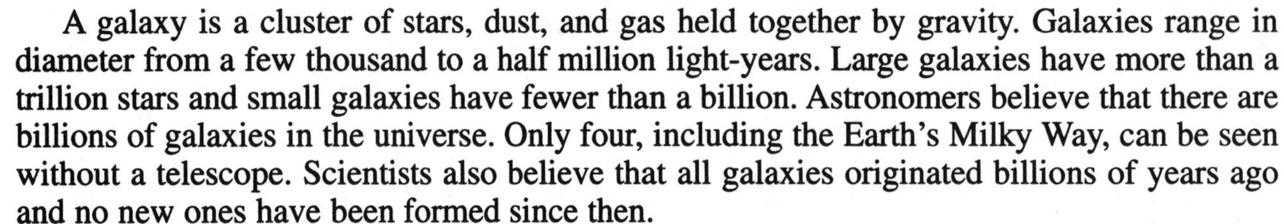

A galaxy is a cluster of stars, dust, and gas held together by gravity. Galaxies range in diameter from a few thousand to a half million light-years. Large galaxies have more than a trillion stars and small galaxies have fewer than a billion. Astronomers believe that there are billions of galaxies in the universe. Only four, including the Earth's Milky Way, can be seen without a telescope. Scientists also believe that all galaxies originated billions of years ago and no new ones have been formed since then.

There are four different galaxy shapes: spiral, barred spiral, elliptical, and irregular. Spiral galaxies resemble a huge pinwheel with spiral arms that coil out from the central mass. Elliptical galaxies range in shape from round to flattened globes. The light from elliptical galaxies is brightest in the center and becomes fainter toward the outer regions.

PROJECT

Construct a mobile that represents the four galaxy shapes.

MATERIALS

- Dark blue or black construction paper
- Galaxy Patterns, following
- Toothbrush
- Coat hanger
- Hole punch
- White tempera paint
- Yarn or string

DIRECTIONS

1. Reproduce the Galaxy Patterns page. Cut the individual patterns apart. Cut along the dotted lines of each to create a stencil.
2. Lay each stencil over construction paper. Dip the bristles of the toothbrush into white paint. Hold the toothbrush over the pattern and gently shake it. The paint should spatter onto the construction paper.
3. Continue to dip and spatter, moving the brush across the pattern area. The heaviest concentration of spattered white paint should be in the center of each stencil.
4. Carefully remove the pattern. Allow the paint to dry.
5. Cut around each galaxy to define its shape. Punch a hole at the top of each. Tie with yarn from a coat hanger to create a mobile.

★★★★ GALAXY PATTERNS ★★★★

MILKY WAY

INFORMATION

The Milky Way is a glowing band of starlight coming from the billions of stars within our own galaxy. At night, we see it as a milky-looking strip of stars because we are inside it. The Milky Way is a spiral galaxy shaped like a disk with a bulge in the center. The stars fan out from the center in wide, curving arms made up of blue-white stars. Older yellow and red stars form the nucleus of the galaxy.

The diameter of the galaxy is about 10 times greater than its thickness. It is so big that light, which travels 186,282 miles (299,791 km) per second, takes about 100,000 light-years to travel from one end to the other.

Our solar system is a tiny speck that is just part of the Milky Way. It is located about 30,000 light-years from the center of the galaxy.

PROJECT

Create a tissue paper collage of the Milky Way.

MATERIALS

- White construction paper
- Dark blue tissue paper
- White, yellow, and red tissue paper
- Starch diluted with water (3:1)
- Paint brush

DIRECTIONS

1. Cut or tear the blue tissue paper into medium-size strips. Place a strip onto the white construction paper. Paint over it with the starch mixture. Add another strip and repeat until the entire surface of the construction paper is covered.
2. Tear or cut the white, red, and yellow tissue paper into a variety of sizes to represent stars and planets.
3. Paint these with starch over the blue background. Recreate the spiral shape as illustrated in the galaxy information. The center (nucleus) should be yellow surrounded by red, then branching off into white arms.

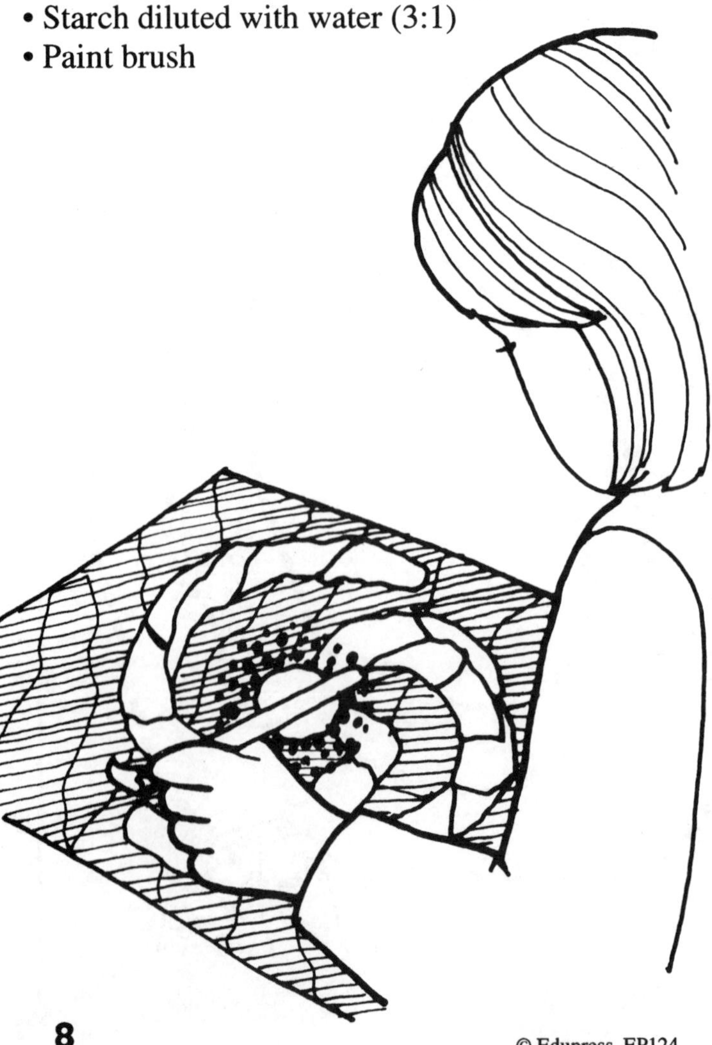

SOLAR SYSTEM

INFORMATION

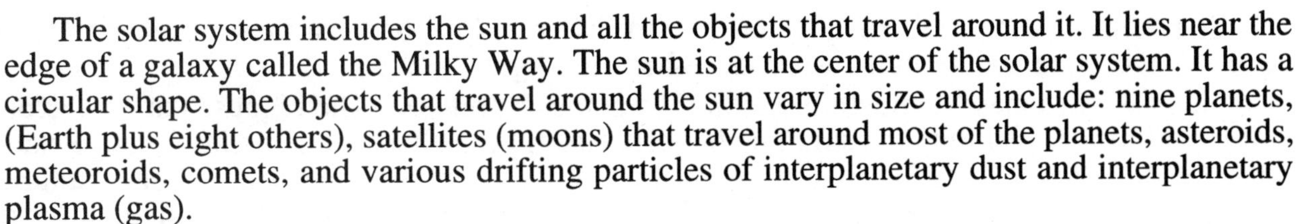

The solar system includes the sun and all the objects that travel around it. It lies near the edge of a galaxy called the Milky Way. The sun is at the center of the solar system. It has a circular shape. The objects that travel around the sun vary in size and include: nine planets, (Earth plus eight others), satellites (moons) that travel around most of the planets, asteroids, meteoroids, comets, and various drifting particles of interplanetary dust and interplanetary plasma (gas).

The solar system is less than one light day across. It is about 30,000 light-years from the center of the Milky Way. Many stars in the Milky Way are the centers of solar systems. The nearest that might have intelligent life is about 100 light years away.

PROJECT

Create a living model of a solar system.

LIVING SOLAR SYSTEM

MATERIALS

- White construction paper
- Marking pens
- 165 people (If you don't have this many, reduce the number of meteoroid, asteroid, comet, dust, and plasma signs. Maintain the ratio.)
- Playground

DIRECTIONS

1. Use marking pens to make construction paper signs in the quantities indicated:
   - 1-sun
   - 9-planet
   - 15-satellite
   - 10-comet
   - 30-asteroid
   - 20-meteoroid
   - 40-dust
   - 40-plasma
2. Invite several classes to participate. Give each person a sign to hold. Go out to the playground and form a very large circle.
3. Call out to the person holding the sun to stand in the center of the circle. Ask the planets to find a position around the sun but at varying distances from it. Position the satellites (moons) around the planets. Fill in with meteoroids and asteroids. Position the comets toward the outside. Scatter the dust and plasma people throughout your living solar system.
4. Discuss the ratio of each solar system member.

STARS

INFORMATION

Stars are glowing balls of gas that live millions and millions of years. A star uses its gases—hydrogen and helium—to produce heat and light.

Stars vary in mass, temperature, color, brightness, and size. The biggest stars are about 100,000 times bigger than the sun. The hottest stars appear blue-white. Yellow-orange stars, like the sun, are cooler. When a star has used up all its hydrogen, the surface cools and turns red. Red stars are the coolest of all. Don't be fooled. Even red stars are incredibly hot—over 5000°F (2800°C).

Except for the sun, most stars are too far from the Earth for their distances to be measured. Of the billions of stars in the universe, only a small percentage can be seen even with the help of the most powerful telescopes.

PROJECT

Create an accordion panel that features three types of stars. Work in cooperative groups to interactively present the autobiography of a star.

MATERIALS

- White construction paper
- Watercolor paints and brush
- Crayons
- Autobiography of a Star idea page, following (reproduce for each cooperative group)
- Resource books

DIRECTIONS

1. Cut a large sheet of white construction paper in half lengthwise. Accordion-fold the length into three six-inch (15.24 cm) sections.
2. Heavily color a circular star in each panel; then paint over it with a watercolor wash to create a crayon resist. Color and paint each star as follows:
   ✳ **First panel:** light blue star; white watercolor resist
   ✳ **Second panel:** yellow star; yellow-orange mix watercolor resist
   ✳ **Third panel:** red star; red watercolor resist

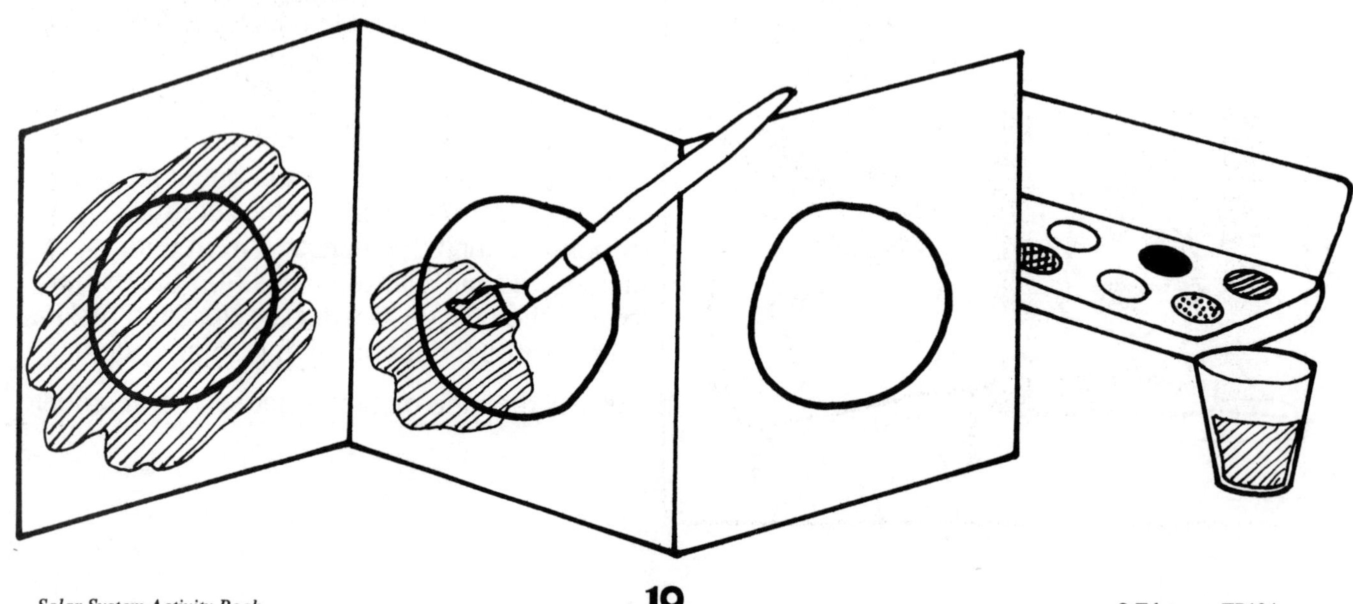

AUTOBIOGRAPHY OF A STAR

A star has an interesting life to share.

Imagine you lived the life of a star. What would you tell others about your birth, your personal traits, your daily life, your death?

Work with some classmates to retell the story of your life. Choose from these ideas or one of your own. Be sure to look through books for facts and information that will help you.

Scrapbook

Create a scrapbook using construction paper and other art materials. Write photo captions to accompany the pictures.

Slide Show

Create picture "slides" from large squares of butcher paper. On each slide present a picture of a different part or your life. Narrate your slide show. Add just the right background music. Present your show, one slide at a time, to classmates.

Short Play

Write a script. Make props and costumes. Assign roles to play. Read your parts aloud and practice your presentation many times. Invite parents, friends, and classmates to your play. Everyone's a star!

CONSTELLATION

INFORMATION

Constellations are groups of stars within a particular region of the sky. By knowing their positions, it is possible to locate stars, planets, comets, and meteors. Throughout the centuries people have used their knowledge of these positions to guide them from place to place during their journeys.

When astronomers in ancient Egypt, Greece, and other lands began to study the sky, they divided it into regions that had distinct groups of stars. 48 of the constellations to which we refer today are the groupings devised by the ancient Greeks. They named the constellations after the shapes they believed they formed.

The stars in a constellation do not necessarily have any relationship to one another. Some may be near the Earth and others far away.

PROJECT

Choose one or all activities and have a constellation celebration.

MATERIALS

- Refer to each activity

STAR SIGNATURES

Each constellation has an identifiable "signature"—the shape it forms in the sky. Learn to recognize these shapes in class. Transfer your knowledge to the nighttime sky.

Materials

- Flashlights

Directions

1. Each star in a constellation is represented by a child holding a flashlight pointed at the ceiling.
2. Stand in the pattern of a constellation. Turn off the overhead lights. Turn on the flashlights. Observe the pattern on the ceiling.
3. Change children and constellation shapes.

CONSTELLATION CELEBRATION

LEGENDS OF THEM ALL

The Early Greeks and Romans developed legends for many of the constellations. Leo the Lion was believed to be the fiercest lion in the world. No weapons would wound him.

Materials

- Writing paper
- Crayons
- Stapler
- Resource books

Directions

1. Fold writing paper in half to form a four-page booklet.
2. Research the ancient legends behind the constellations. Choose four to retell and illustrate in a journal. Share the legends with classmates.

STAR STRUCK

For mapping purposes, astronomers divide the sky into 88 constellations.

Materials

- Star stickers
- Paper
- Resource books

Directions

1. Use star stickers to recreate as many constellations as you can find in your research.
2. Have a constellation contest to find out who can make and identify the most.

MINI MAPS OF THE STARS

Materials

- Writing paper
- Crayons
- Stapler
- Resource books
- Straight pin

Directions

1. Reproduce the Constellation Pattern page, following.
2. Staple the pattern page to black construction paper.
3. Working against a soft carpeted surface, poke the straight pin through all the lines connecting the stars. Each hole should be very close to the one next to it. Several holes should be made at the location of a star.
4. When all lines are complete, turn the paper over. Looking through the black side, hold the paper up to the light to reveal mini-constellations.

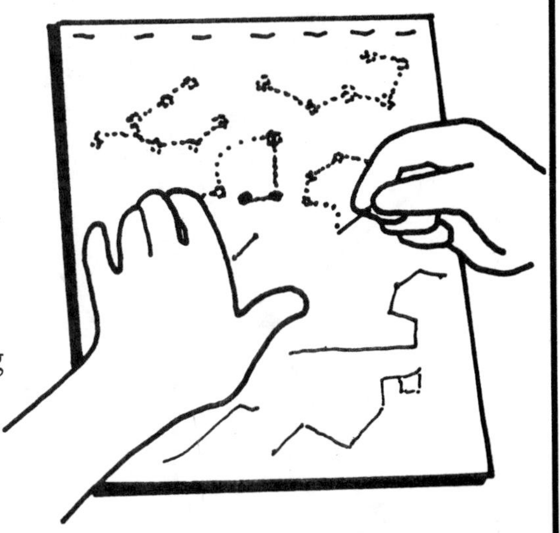

CONSTELLATION PATTERNS

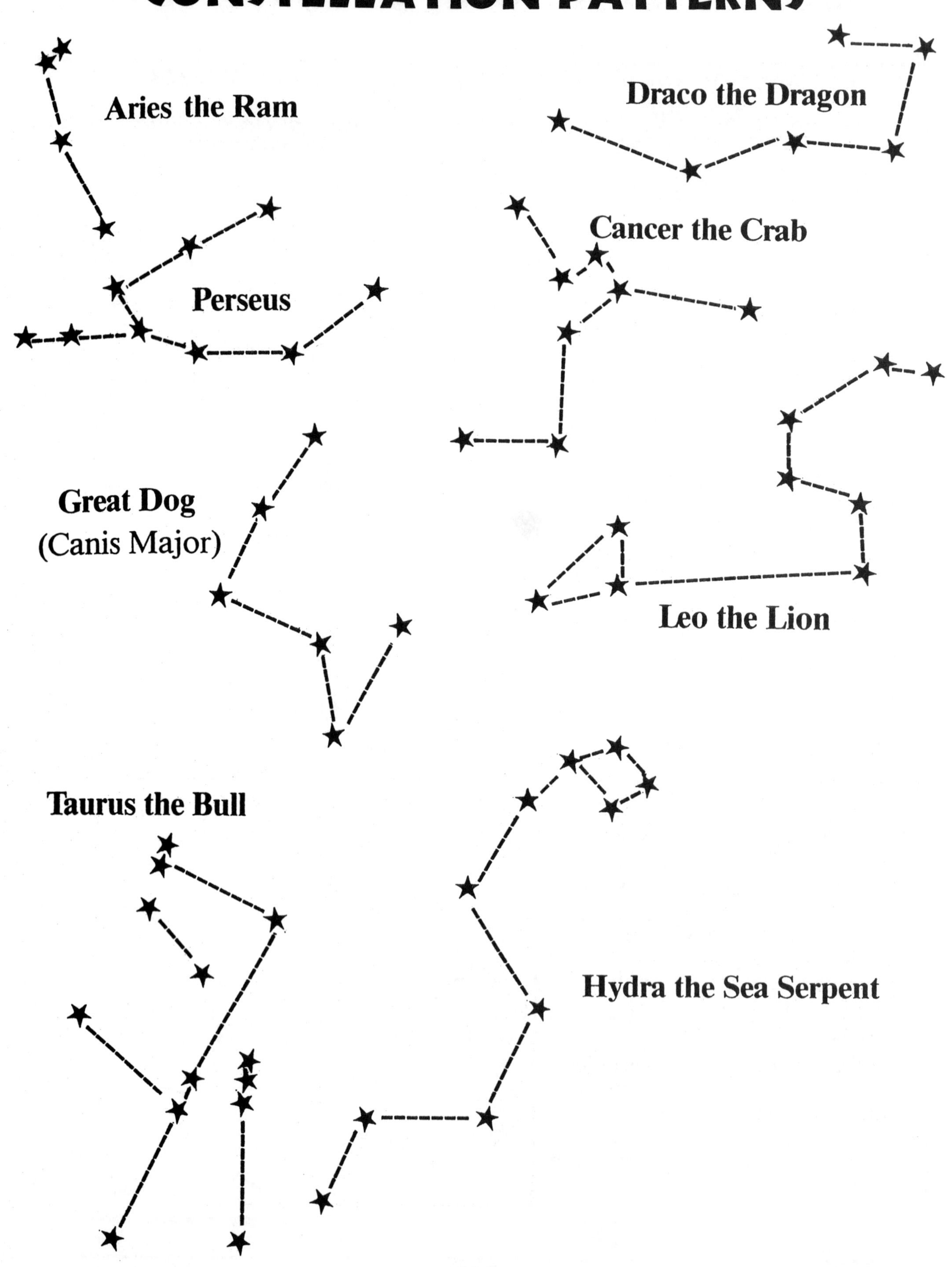

STAR NAMING

INFORMATION

Most star names are derived from names in Greek, Roman, and Arabic mythology. Professional astronomers use a different method. A constellation is identified by its Latin name. Within each constellation the stars are identified by a Greek letter of the alphabet in order of brightness. The brightest star in a constellation is Alpha, the next brightest is Beta, and so on. The brightest star in Orion is known as Alpha Orionis. However, there are only 24 letters in the Greek alphabet and there are 88 constellations, so only 2,112 stars are identified using this system. The remaining millions are identified by astronomical coordinates or catalog numbers. The Guide Star Catalogue, created for the Hubble orbiting telescope, identifies over 15 million stars. Because there are millions of stars with no name, it is possible to name a star through the International Star Registry.

PROJECT

Participate in a star naming and mapping activity.

MATERIALS

- Construction paper
- Crayons
- Pencil
- Paper plate
- Marking pens
- Stapler

DIRECTIONS

1. Use marking pens to make a simple map of the universe. Include the Milky Way and our solar system within it. Identify planets and known stars.
2. Locate your personal star within the universe. Indicate which one it is by using a different color marking pen.
4. Using information about stars and the solar system, write about your star on a paper plate. Include:
   - Its name and the reason for the name
   - Its location in the universe
   - The number of light-years away
   - Its age
   - Its size
5. Color the plate based on the age of the star (see page 11). Staple the plate to the map and display.

EXAMPLES

Since 1978, people have been naming stars for themselves, their loved ones, their heroes, or with an abstract concept. Some star names:

★ *Famous people*—Sylvester Stallone
★ *Couples*—Rosemary and Howard
★ *Ideas*—Dreams Come True in Time

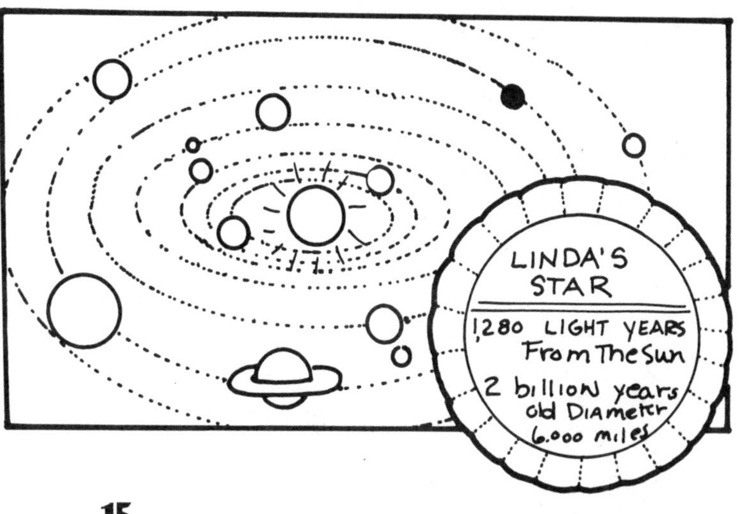

SUN

INFORMATION

The sun is only one of billions of stars in the universe. It's about average in size and temperature compared to other stars. At about 4½ billion years of age, it has lived about half its life. It is made mainly of hydrogen and helium.

The sun is the center of our solar system. Earth and eight other planets travel around it. If you put together all the planets, moons, comets, and other bodies in the solar system, the sun would still be 500 times bigger. It is the closest star to Earth. Like Earth, the sun rotates on its axis. This rotation takes about a month.

The sun's energy is generated at its center where the temperature reaches nearly 30,000,000°F (14,888,000°C). When the energy reaches the surface it is sent into space in the form of heat and light. Without the heat and light of the sun there would be no life on Earth.

PROJECT

Draw a diagram of the sun's interior. Conduct simple experiments that demonstrate the effects of the sun's energy.

MATERIALS

- Butcher paper
- Pencils
- Marking pens
- Tempera paints and brushes
- Resource books

DIRECTIONS

1. On butcher paper, draw a large outline of the sun.
2. Use information from the diagram at right, as well as research, to label and explain important facts about our sun.
3. Share and compare the diagrams and information.
4. Brainstorm a cooperative list of facts you learned about the sun.

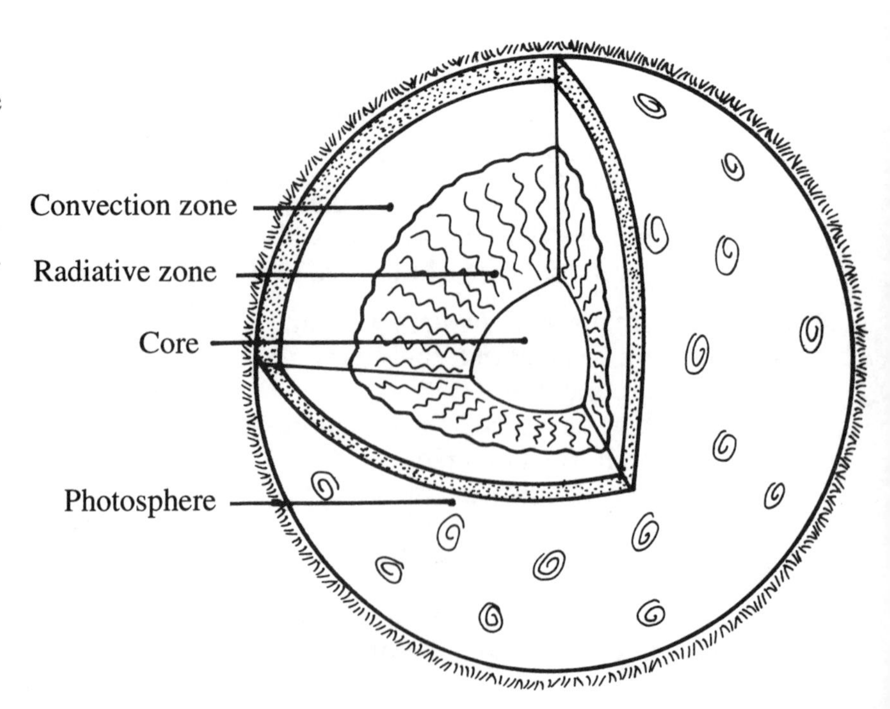

SUN EXPERIMENTS

SOLAR PRINTS

Materials

- Black construction paper
- Assorted objects such as rubber bands, pencils, erasers, scraps of paper, scissors

Directions

1. Start this experiment early in the morning on a bright, sunny day.
2. Set the black construction paper outside. Choose five to seven objects of varying sizes to set on the paper.
3. Toward the end of the day, remove the objects from the paper and examine the results. What causes the paper to fade around the objects?
4. Display the results on a classroom bulletin board.

SOLAR HEAT

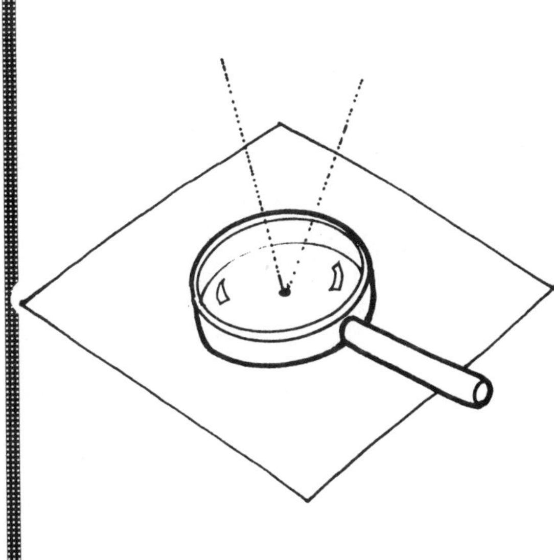

Materials

- Magnifying glass
- Piece of paper

Directions

1. Use the magnifying glass to focus sunlight on a piece of paper on the ground. The light is focused when the dot of sunlight is perfectly round and as small as possible. (Don't look directly at the light.)
2. After a while, touch the paper. Do you feel a change in temperature?

PLANETS

INFORMATION

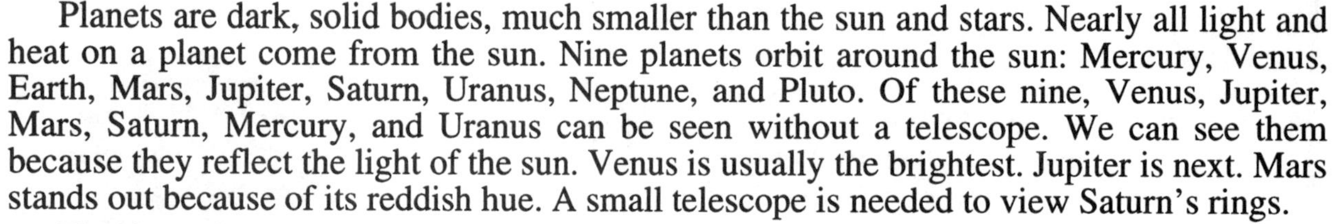

Planets are dark, solid bodies, much smaller than the sun and stars. Nearly all light and heat on a planet come from the sun. Nine planets orbit around the sun: Mercury, Venus, Earth, Mars, Jupiter, Saturn, Uranus, Neptune, and Pluto. Of these nine, Venus, Jupiter, Mars, Saturn, Mercury, and Uranus can be seen without a telescope. We can see them because they reflect the light of the sun. Venus is usually the brightest. Jupiter is next. Mars stands out because of its reddish hue. A small telescope is needed to view Saturn's rings.

The four planets closest to the sun—Mercury, Venus, Earth, and Mars—are rocky and small. The next four from the sun—Jupiter, Saturn, Uranus, and Neptune—are gaseous and large. Jupiter, Saturn, and Uranus are surrounded by rings. The outermost planet—Pluto—is comparatively small. All the planets except Mercury and Venus have one or more moons.

PROJECT

Create an interactive planet-fact bulletin board to use as a springboard for classroom games.

MATERIALS

- Planet Fact Cards, following
- Nine sheets of construction paper
- Pencils
- Shoe box
- Marking pens
- Masking tape

DIRECTIONS

1. Write the name of a planet on each sheet of white construction paper.
2. Tape them to the classroom wall in a row to create a large bulletin board. Allow room around each planet name.
3. Reproduce the Planet Fact Cards. Cut apart the cards and put them in a decorated box near the bulletin board. Include scissors, tape, and pencils in the box.
4. Challenge students to search for planet facts, filling in a fact card and taping it to the bulletin board below the matching planet. Encourage them to constantly review the facts on the wall so as to not duplicate.
5. Use the facts as the basis for classroom games such as "Name that Planet" or "Sixty Second Fact Race."

PLANET FACT CARDS

PLANET ____________________
FACT ____________________
____
Fact Finder ____________________

PLANET ____________________
FACT ____________________
____
Fact Finder ____________________

PLANET ____________________
FACT ____________________
____
Fact Finder ____________________

PLANET ____________________
FACT ____________________
____
Fact Finder ____________________

PLANET ____________________
FACT ____________________
____
Fact Finder ____________________

PLANET ____________________
FACT ____________________
____
Fact Finder ____________________

PLANET ____________________
FACT ____________________
____
Fact Finder ____________________

PLANET ____________________
FACT ____________________
____
Fact Finder ____________________

PLANET ____________________
FACT ____________________
____
Fact Finder ____________________

PLANET ____________________
FACT ____________________
____
Fact Finder ____________________

PLANET COMPARISONS

INFORMATION 

Planets have certain features in common. Each rotates around the sun. Each spins around a rotational axis, an imaginary line through its center, as it circles the sun. All have an atmosphere made up of a mixture of gases that surround it.

All planets vary tremendously in size. Jupiter, the largest planet, is 45 times as large as Pluto, the smallest.

Other conditions on the planets, including temperature, atmosphere, surface features, length of days and nights, are very different. Much of this difference can be attributed to the planet's distance from the sun. Astronomers are able to determine all this information by studying the light, radio waves, and other radiation coming from the planet.

PROJECT

Conduct research to complete a planet comparison chart. Create a visual display that compares the different sizes of the planets.

MATERIALS

- Planet Comparison Chart, following
- Resource books, encyclopedias
- Pencils
- Ten objects—basketball, softball, baseball, golf ball, two ping-pong balls, marble, small marshmallow, kernel of unpopped corn

DIRECTIONS

1. Divide students into pairs. Reproduce the Planet Comparison Chart for each pair.
2. Review the comparison chart. Discuss the kind of information each column is asking for and where this information can be found.
3. Set a due date for completion of the chart. Compare information. What might account for a variation in answers?
4. Using information from the charts, place the objects listed above to create a visual display of the planets according to their size and relationship to the sun. Start with the basketball (the sun) in the center.

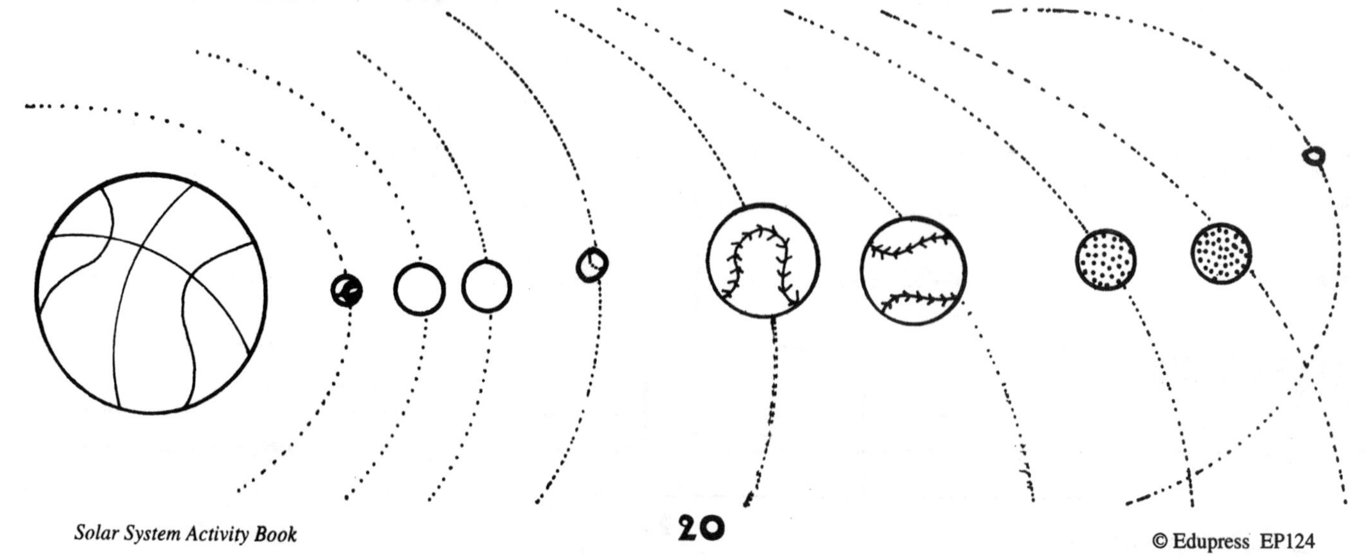

PLANET COMPARISON CHART

	Distance from the sun	Diameter	Number of Satellites	Period of Rotation	Orbital Speed
EARTH					
URANUS					
NEPTUNE					
PLUTO					
MERCURY					
VENUS					
SATURN					
JUPITER					

ORBIT

INFORMATION

An orbit is the path of any object whose motion is controlled by the gravitational pull of another object. Planets, asteroids, meteoroids, and comets orbit the sun. Satellites (moons) orbit planets. Planetary orbits are somewhat oval in shape. This shape is called an *ellipse,*

Rotation and *revolution.* These are key words to understanding the movement of planetary orbits. A planet spinning on its axis rotates. A single rotation equals one day. One trip around the sun is called a revolution and equals one planetary year. Planets revolve and rotate at different speeds. The planets closest to the sun move around their orbits faster than those farther away. For example, Earth rotates once every 24 hours. Jupiter rotates once every ten hours. But Earth's year is 365 days. Jupiter's is 4,329 days. Jupiter may rotate faster than Earth, but one revolution takes over 11 Earth years!

PROJECT

Play a game to gain an understanding of the concepts of rotation, revolution, and orbit.

MATERIALS

- Playground or large room clear of furniture
- Several basketballs or balls of similar size

DIRECTIONS

1. First learn the difference between rotation and revolution. Spin a ball in place. This is rotation. Hold a ball and walk in a circle. This is revolution. Spin a ball in the palm of your hand while walking in a circle. This is an orbit.
2. Sit down about three feet (1 m) apart to form an oval around a center object, the sun. Spin a ball while rolling it around the circle from person to person.
3. Form a larger circle around the first one. At a signal, spin and move the ball around each circle. Add more circles and repeat the orbit activity. What observations can you make? Experiment with different rotation and revolution speeds.

ATMOSPHERE

INFORMATION

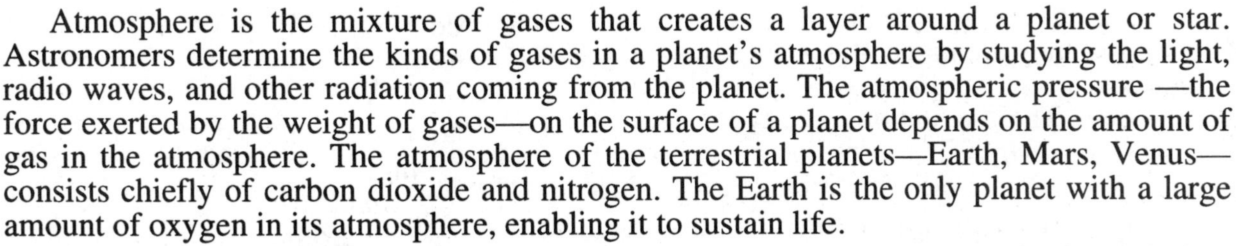

Atmosphere is the mixture of gases that creates a layer around a planet or star. Astronomers determine the kinds of gases in a planet's atmosphere by studying the light, radio waves, and other radiation coming from the planet. The atmospheric pressure —the force exerted by the weight of gases—on the surface of a planet depends on the amount of gas in the atmosphere. The atmosphere of the terrestrial planets—Earth, Mars, Venus—consists chiefly of carbon dioxide and nitrogen. The Earth is the only planet with a large amount of oxygen in its atmosphere, enabling it to sustain life.

The atmosphere around a planet usually determines the color it appears in the nighttime sky. The chart below provides more information about each planet and its atmospheric color.

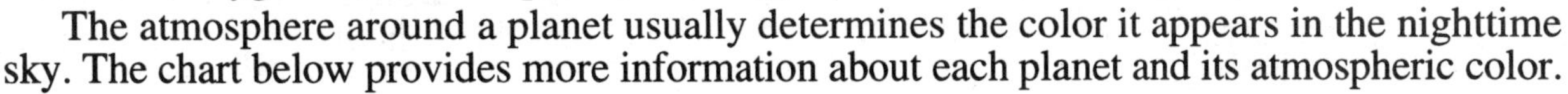

PROJECT

Create an informational mural of the planets.

MATERIALS

- Dark blue or black butcher paper
- Scissors
- Glue
- Index cards
- Construction paper
- Crayons, paints
- Resource books

DIRECTIONS

1. Tape a large piece of butcher paper to the classroom wall.
2. Divide into nine small groups. Assign each group a planet for atmospheric research.
3. Use paper and art medium of choice to duplicate the color of each planet. Report to classmates on the atmospheric make-up of the assigned planet. Describe the planetary conditions created by the presence of these gases.
4. Work together to glue the planets to the mural in their correct solar system position in relationship to the sun. Display the atmospheric gases on index cards glued below each planet.

Mercury	dark blue
Mars	red
Venus	orange yellow
Jupiter	pale yellow
Saturn	golden brown
Earth	blue/white swirls
Uranus	pale green/blue
Neptune	aqua
Pluto	white

GRAVITY

INFORMATION

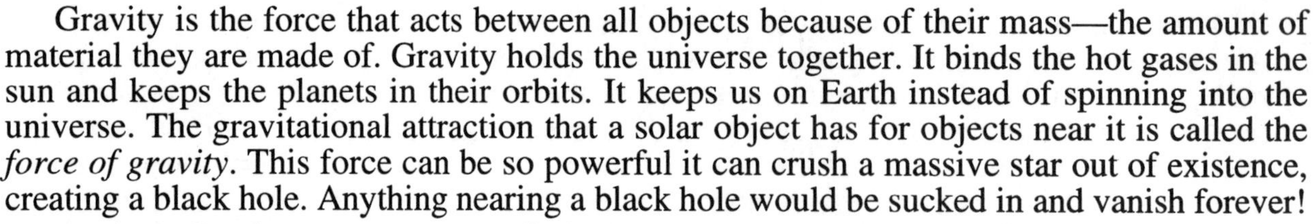

Gravity is the force that acts between all objects because of their mass—the amount of material they are made of. Gravity holds the universe together. It binds the hot gases in the sun and keeps the planets in their orbits. It keeps us on Earth instead of spinning into the universe. The gravitational attraction that a solar object has for objects near it is called the *force of gravity*. This force can be so powerful it can crush a massive star out of existence, creating a black hole. Anything nearing a black hole would be sucked in and vanish forever!

Ancient Greek astronomers studied planetary movement. But it wasn't until the late 1600s that an English scientist, Isaac Newton, observed an apple falling from a tree and went on to conclude that there is a connection between the force that attracts objects to the Earth and the way the planets move.

PROJECT

Use the scientific method to conduct gravity experiments.

MATERIALS

- As listed for each experiment

BLACK HOLE

- **Hypothesis:** What effect does the gravitational pull of a black hole have on objects around it in space?
- **Materials:** Vacuum with hose; objects of various sizes
- **Method:** Remove all hose attachments. Place objects at varying distances from the hose. Turn on the vacuum. Observe the movement of the objects. Repeat the experiment several times changing the position of the objects. Time movement, measure distances, draw conclusions.

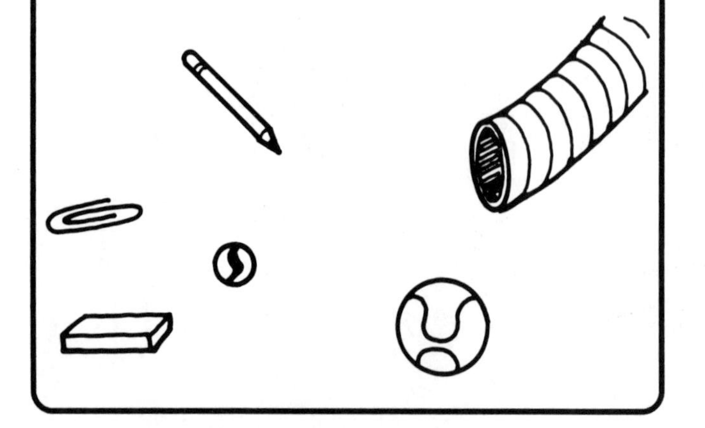

BLACK HOLE

- **Hypothesis:** What effect does mass (weight) of an object have in relation to gravity?
- **Materials:** Scale; unbreakable objects of various weight; tape measure; stopwatch
- **Method:** Record the weight of each object. Make a mark on the wall at least six feet (1.8 m) above the floor. Drop an object from the mark. Time its descent until it hits the floor. Conduct gravity races in which two objects are dropped at the same time. Draw conclusions.

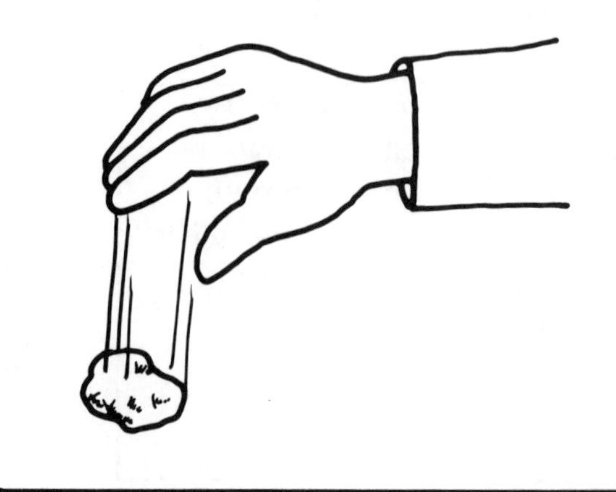

MERCURY

INFORMATION

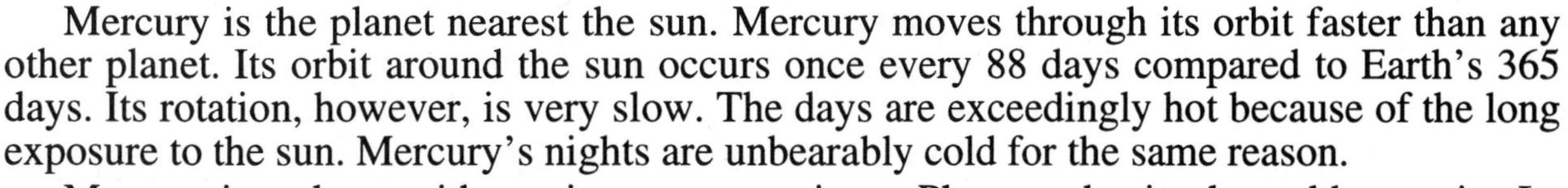

Mercury is the planet nearest the sun. Mercury moves through its orbit faster than any other planet. Its orbit around the sun occurs once every 88 days compared to Earth's 365 days. Its rotation, however, is very slow. The days are exceedingly hot because of the long exposure to the sun. Mercury's nights are unbearably cold for the same reason.

Mercury is a planet with no air, oceans, or rivers. Plants and animals could not exist. Its lack of atmosphere allows the planet to go unprotected against crashing meteoroids and asteroids. Consequently, the surface of Mercury is covered with thousands of craters. Between the craters are smooth plains and huge slopes. Its pocked surface hides a core twice as iron-rich as any other planet. Scientists are still unsure as to whether or not this core is liquid or solid.

PROJECT

Gain an understanding of Mercury's mass with an eating activity.

MATERIALS

- Oranges
- Plastic knives

DIRECTIONS

1. Cut an orange in half. Examine the layers.
2. Imagine the rind to be the crater-marked surface of Mercury.
3. Just below the surface is the crust—the white pithy part of an orange rind.
4. The center fruit is the core.
5. Go ahead. Eat Mercury. If you are a scientist who believes the planet's core is solid, eat the orange section by section. If you are a scientist who believes the core is liquid, squeeze the juice from the sections.

VENUS

Venus is the second closest planet to the sun. Its size and mass are similar to Earth. Venus is shrouded by a yellowish cloud cover of poisonous sulfuric acid. The reflection of the sun off these clouds makes Venus the brightest object in the sky apart from the sun and the moon. Beneath this cloudy atmosphere, lightning flashes day and night across a rainless sky. Thunder booms continuously. The pressure from the thick atmosphere would make you feel as though you were 300 feet (91.4 m) under water. The speed of Venus' orbit is second only to Mercury, but its rotation is very slow. Venus circles the sun every 224 days.

Its surface of fine ash or sand is brutal. Temperatures reach 880°F (300°C). Visibility is poor. Venus' landscape features great canyons, rugged mountain ranges, shallow craters, lakes of molten lava, and vast, flat plains.

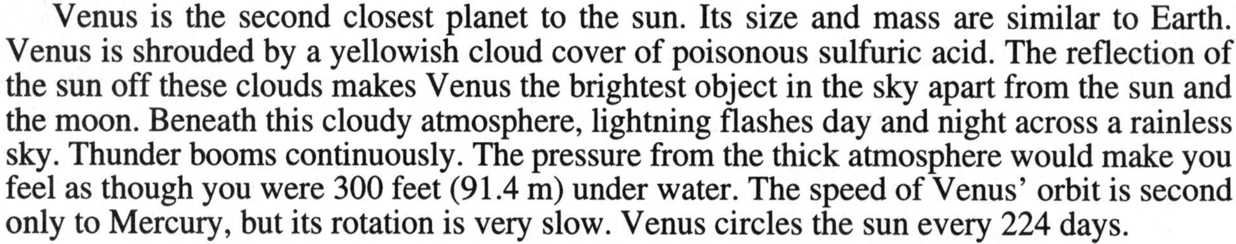

PROJECT

Create a sensory experience that simulates a visit to Venus.

MATERIALS

- Several electric fans
- Cymbals, drums, pan lids
- Pan of water, hot plate
- Eggs
- Flashlights
- Darkened room

DIRECTIONS

1. Set a pot of water and eggs on the hot plate and bring it to a boil. When the smell of sulphur is evident, continue the experience. (Egg boiling is optional but it does involve the sense of smell in the experience.)
2. Arrange several electric fans around the room. Give half the students flashlights and the other half cymbals, drums, or pan lids.
3. Turn off the classroom lights and begin the thunder and light storm. Turn on all the fans. Bang the lids, cymbals, and drums. Flash the flashlights as fast as you can and flip the room lights on and off.
4. Be thankful your trip did not include feeling the touch of Venus' poisonous acid rain, the weight of its atmosphere or the intensity of its surface heat!

EARTH

INFORMATION

Earth was formed more than four billion years ago. A colorful planet of green spaces, deserts, deep oceans, and fields of ice, 70 percent of its surface is covered by water. It is the only planet where there is life—made possible because Earth is in a zone called the "ecosphere"—just the right distance from the sun. Closer, it would be too hot to survive; farther away, it would freeze. All life on Earth is found on and above a skin-like crust made of rock. Beneath the crust is a hot, lifeless ball of rock and metal.

Earth rotates once every 24 hours—one day. It takes 365 days to complete an orbit around the sun—one year. Seasons are caused by the way Earth tilts as it orbits the sun. Throughout the year, the southern and the northern hemispheres have opposite seasons.

Earth has one natural satellite, the moon, held in place by Earth's gravitational pull.

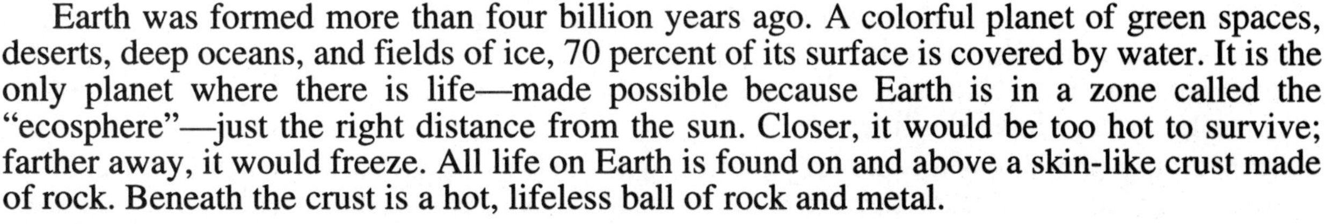

PROJECT

Conduct an experiment relating to the ecosphere.

MATERIALS

- Three small potted plants
- Sun lamp
- Freezer

ECOSPHERE EXPLORATION

▼ **Hypothesis:** What would happen to life if Earth were closer to or farther from the sun?

▼ **Method:** Place each plant in a different environment—the freezer; directly under a sun lamp; a well-ventilated place in the classroom.

▼ **Results:** Record your observations for several days.

▼ **Conclusions:** What comparisons can be drawn between the results and Earth's ecosphere?

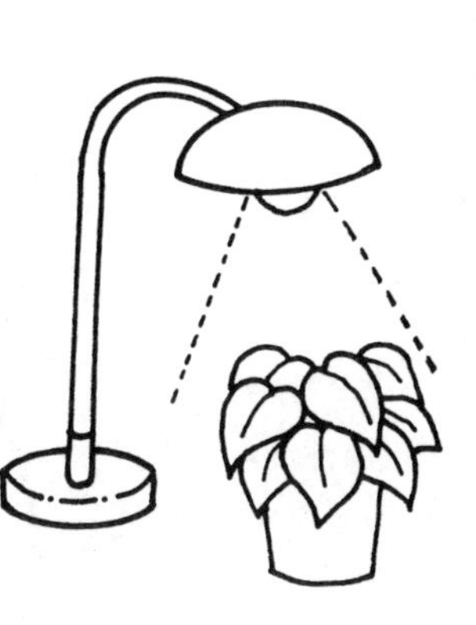

SATELLITE

INFORMATION

A satellite is a natural body that orbits about a larger body in space. The word satellite means attendant. We can think of the satellites that orbit the planets in our solar system as attendants to them. Natural satellites are sometimes called moons. At least seven planets in our solar system have satellites.

Mercury and Venus are the only planets without any satellites. Each satellite has its own unique physical properties and theory as to how it was created. It is usually given a name or numerical identification. Satellites are continually being discovered by astronomers. Check the copyright year of your reference material for the most current information.

The faint or dark bodies that revolve about certain stars and cause their light to dim and brighten are also called satellites. The stars they attend are called eclipsing or variable stars.

PROJECT

Graph the planetary satellites.

MATERIALS

- Graphing paper
- Crayons
- Pencil or pen
- Resource books

DIRECTIONS

1. Divide the graphing paper into nine columns. Label each column with a planet name.
2. Research (or use the chart below) to determine the number of satellites each planet has. Create a graph illustrating this information.
3. Conduct additional research to find out more about the satellites that attend the planets.

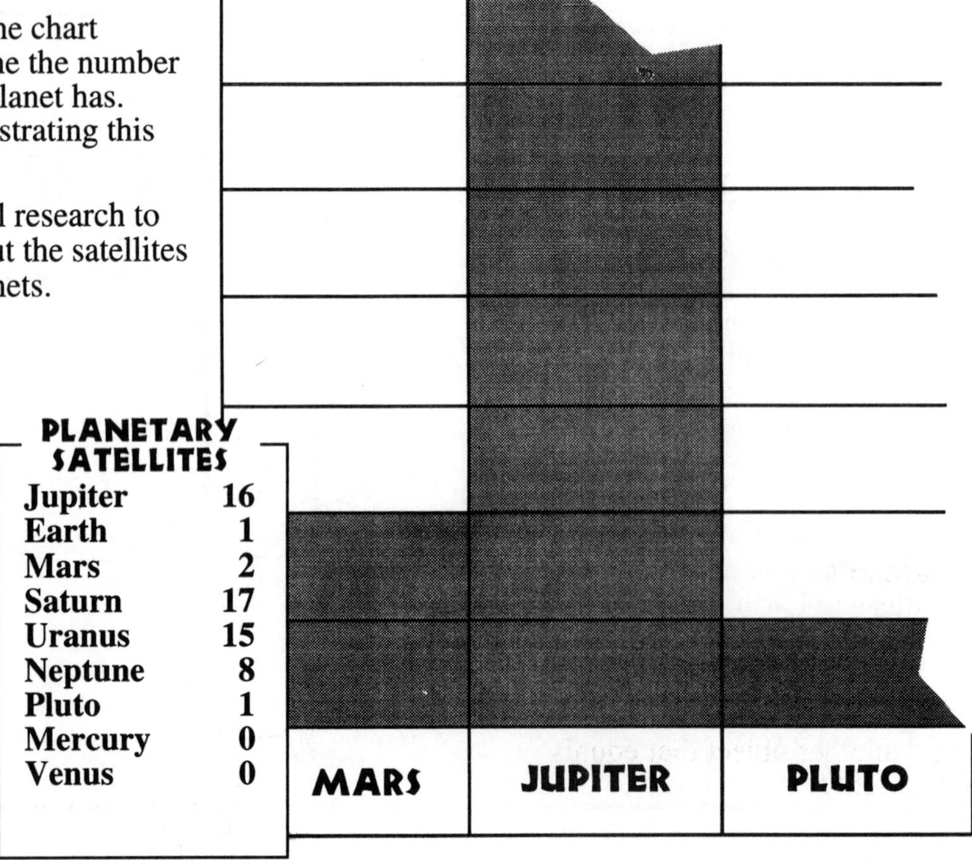

PLANETARY SATELLITES

Planet	Satellites
Jupiter	16
Earth	1
Mars	2
Saturn	17
Uranus	15
Neptune	8
Pluto	1
Mercury	0
Venus	0

EARTH'S MOON

INFORMATION

The moon is Earth's only natural satellite and its nearest neighbor in space. The moon travels around Earth about once every 29.5 days. The pull of Earth's gravity keeps the moon in its orbit. In turn, the moon's gravity pulls on Earth's oceans, causing high and low tides. Its gravity is much weaker than Earth's so you could jump almost six times higher if you were on the moon.

The moon gives off no light of its own. When it shines, it is reflecting light from the sun. The moon has no life of any kind. It has no air, no wind, and no water. On the moon, the sky is black even during the day.

The surface of the moon is marked with craters. There are lowlands of rock covered by a thin layer of rocky soil. There are highlands that are rough and mountainous. Its soil consists of ground rock and bits of glass.

PROJECT

Perform some math calculations related to gravity on the moon.

MATERIALS

- Measuring tape
- Butcher paper
- Crayon
- Scale
- Scratch paper
- Calculator (optional)

MOON JUMP

1. Tape a large sheet of butcher paper to the wall. Stand with your shoulder at a right angle against the wall. With your arm closest to the wall, hold onto a crayon and raise your arm straight in the air.
2. Bend and jump as high as you can. Make a mark with the crayon on the butcher paper at the height of your jump.
3. Measure the distance from the floor to the mark. Multiply by six to calculate your jumping height on the moon.

MOON WEIGHT

1. Weigh yourself on a standard scale. Record the weight. Divide by six. That is your weight if you were on the moon.
2. Can you find another object that equals your moon weight?

THE SOLAR SKY

INFORMATION

The moon is the brightest object in the nighttime sky. Its phases—full, half, and crescent—occur because the amount of its sunlit area that can be seen from the Earth changes as it orbits the Earth. The moon goes through a complete set of phases about once a month. More stars and planets are visible on a dark, moonless night. The planets appear first. Stars come out when the sky is dark. People who live near the equator can see all the stars in the sky during the course of a year.

As sunlight passes through the Earth's atmosphere, it strikes molecules of the gases that make up the atmosphere. The sky becomes so bright that stars and planets cannot be seen during the day. Only the moon is bright enough to sometimes be faintly visible.

PROJECT

Maintain an illustrated journal of the daytime and nighttime solar sky.

MATERIALS

- White drawing paper
- Pencils
- Calendar
- Stapler
- Crayons

DIRECTIONS

1. Fold several pieces of drawing paper in half. Staple at the fold to create a booklet.
2. Draw a horizontal line to divide each page in half. Label the top half daytime. Label the bottom half nighttime. Select a one-month period during which students will observe the daytime and nighttime sky. Record observations with illustrations and short written descriptions.
3. Note the date at the top of each page.

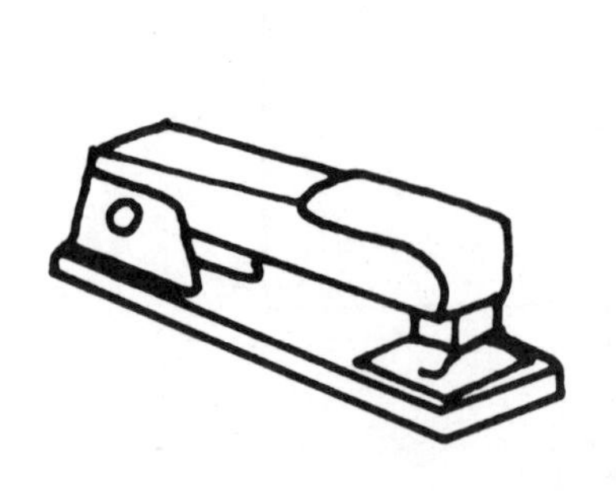

MARS

INFORMATION

Mars is the next planet from the sun after the Earth. A day on Mars lasts almost 25 hours. Its year is 687 days. Two tiny, heavily-cratered moons orbit the planet. The atmosphere is too thin to breathe so there is no animal or plant life on this freezing cold planet.

The surface of Mars has giant volcanoes that are often splashed with snow. Long ago they erupted, but today they are all dead. The biggest volcano is Olympus Mons. It is three times bigger than the highest mountain on Earth but so wide that the incline to its peak is not steep. Mars also has huge canyons and mountains. There are red rocks and rusty iron dust everywhere. There are ice caps on its north and south poles. Dust storms created by strong winds can last for weeks. Even the sky around the planet is red.

PROJECT

Create a diorama that features the surface of Mars.

MATERIALS

- Shoe box
- Red paint
- Red tissue paper
- Paint brush
- Glue
- Scissors
- Red food coloring **or** red tempera paint
- Rocks and sand

DIRECTIONS

1. Take the lid off the shoe box. Paint the inside of the box red. Allow to dry.
2. Stand the box on its long side. Use a variety of art materials to create Mars' surface. Include the features related in the information section.
3. Red food coloring or tempera paint may be used to color sand a reddish hue.

ECLIPSE

INFORMATION

An eclipse occurs when the shadow of one object in space falls on another object or when an object moves in front of another to block its light.

An eclipse of the sun (solar eclipse) occurs when the moon passes in front of the sun and blocks the light to places on the Earth's surface. Solar eclipses can be *partial,* when only a part of the sun is blocked from the Earth's view; *total,* when the sun is totally blocked from the Earth's view; or *annular,* when the sun's light is still visible around the edge of the moon. It is dangerous to look directly at the sun during an eclipse!

An eclipse of the moon (lunar eclipse) occurs when the shadow of the Earth darkens the moon. Lunar eclipses can be either partial, when the moon is only partly covered by the Earth's shadow, or total, when the moon is totally covered by the Earth's shadow.

PROJECT

Demonstrate the difference between a lunar and solar eclipse.

MATERIALS

- Large spotlight or work light
- Two butcher paper circles with these diameters:

6 inches	(15.24 cm)	moon
36 inches	(91.44 cm)	Earth

- Yarn
- Scissors
- Large paper clip

DIRECTIONS

1. Cut circles the size indicated. Remember that the Earth is about ten times the size of the moon.
2. Tie a paper clip to the end of a 4-foot (1.22-m) length of yarn. Hang the yarn from the ceiling.
3. Set the spotlight in the center of the room. Position it so it is directly opposite the paper clip.
4. Clip the Earth to the yarn. Tape the moon to the wall directly behind the Earth. Observe the results. Reverse the positions. What role does size play in the shadowing effect of an eclipse?

Just for fun: Try some human eclipses! Can one student cover another with his or her shadow? *Caution*: Don't look directly into the spotlight.

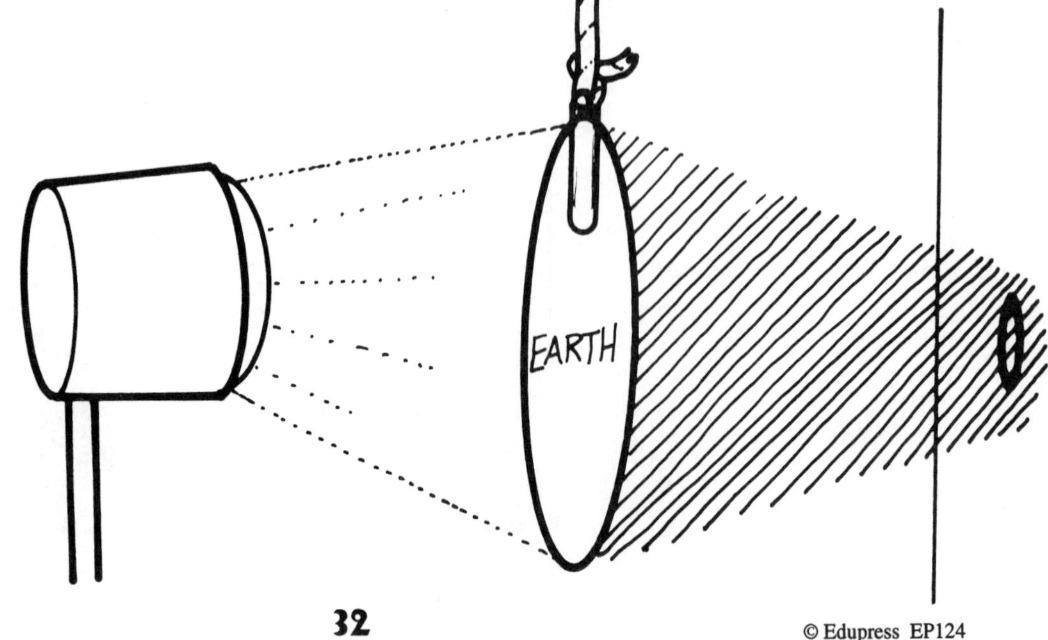

JUPITER

INFORMATION

Jupiter is the largest planet in the solar system and the fifth closest to the sun. It is one and a half times larger than all the other planets put together. Jupiter takes almost 12 Earth-years to orbit the sun and rotates faster than any other planet. Jupiter has 16 known satellites.

Astronomers believe that Jupiter is a huge ball of gas. An icy frosting of clouds several layers thick covers what most of Jupiter is made of—hot liquid hydrogen. Its atmosphere is a churning sea of clouds of many colors. A giant storm, referred to by astronomers as the Great Red Spot, has raged for at least 300 years generating winds up to 224 miles (360 km) per hour. Lightning fills the upper atmosphere. Jupiter has a greater mass than any other planet—318 times as massive as the Earth. The force of gravity is greater as well. In spite of its mass, Jupiter is only slightly denser than water.

PROJECT

Recreate Jupiter's colorful atmosphere with a swirl painting.

MATERIALS

- Tempera paint
- Large sheet of newsprint or white butcher paper
- Plastic spoons
- Scissors

DIRECTIONS

1. Fold the newsprint in half. Reopen and lay the newsprint flat on a table top.
2. Drop a large spoonful of red paint near the fold on the lower third of the newsprint.
3. Drop five to seven more paint blobs of varying colors around the central area of the newsprint taking care to make no blob larger than the red one.
4. Refold the newsprint. Use the palm of your hand to press down and outward to move and blend the paint.
5. Open the newsprint. Allow the paint to dry. Hopefully the red area is most prominent. Cut a very large circle around the painted area. Feature colorful Jupiter on a bulletin board. The Great Red Spot should lie in the "southern hemisphere" of the planet painting.

SATURN

INFORMATION

Saturn is the second largest planet and the sixth closest planet to the sun. Saturn has 95 times more mass than Earth but it takes up about 800 times as much room. It is more liquid than solid. Saturn's density is 70% that of water. It is the least dense of all the planets.

Saturn is surrounded by seven thin, flat rings at its equator and a layer of liquid hydrogen that creates gold-colored clouds. The rings do not touch the planet. The narrow ringlets are made up of billions of pieces of ice ranging in size from dust particles to chunks that measure more than ten feet (3 m) in diameter. Saturn's outermost ring may measure as wide as 180,000 miles (289,674 km). If you could take a walk around the outer edge of the outside ring going 15.5 miles (25 km) a day you wouldn't be back at your starting place for 95 years!

Astronomers believe Saturn has as many as 18 moons, the most in the solar system.

PROJECT

Make a solar treat and munch on the icy particles of Saturn's rings.

MATERIALS

- Blender
- Ice cubes
- Paper cups
- Pineapple juice
- Plastic spoons

DIRECTIONS

1. Crush ice in a blender.
2. Spoon into paper cups. Pour pineapple juice to simulate the yellow rings of Saturn.
3. Munch on Saturn's rings and share facts about this giant planet.

URANUS

INFORMATION

Uranus is the seventh planet from the sun and nearly four times the size of Earth. The third largest of the four gas planets, Uranus is made up mainly of hydrogen and helium. Gas and ice form so much of the planet its gravity is very weak.

The axes of most planets in the solar system are perpendicular to the sun, but Uranus is tilted completely on its side, perhaps the result of a collision with another solar body. It orbits the sun every 84 years. This means that each pole has constant sunlight for 42 years.

Dense clouds cover Uranus. Its temperature is estimated to be -357°F (-216°C). At least nine narrow rings, discovered by astronomers in 1977, surround it. The rings appear to be composed of lumps of dark rock. Five moons circle the planet. Uranus shines with a pale greenish-blue color but is so faint it is easily overlooked in the nighttime sky.

PROJECT

Demonstrate the tilt of Uranus' axis compared to other planets.

MATERIALS

- Cardboard
- Scissors
- Compass
- Two rubber playground balls
- Measuring tape

DIRECTIONS

1. Measure the diameter of the balls. Use the compass to draw a circle the same diameter on the cardboard. Draw a second circle around the first that is two inches (5.08 cm) larger in diameter.
2. Cut around the outside circle. Cut out the inside of the inner circle to create a cardboard ring. Use this ring as a pattern to make a second ring.
3. Place each ring around a rubber ball. Demonstrate the position of Uranus' axis as it orbits by rolling it on its ring. Demonstrate the orbiting axis of the other planets by spinning the second ball, as shown.

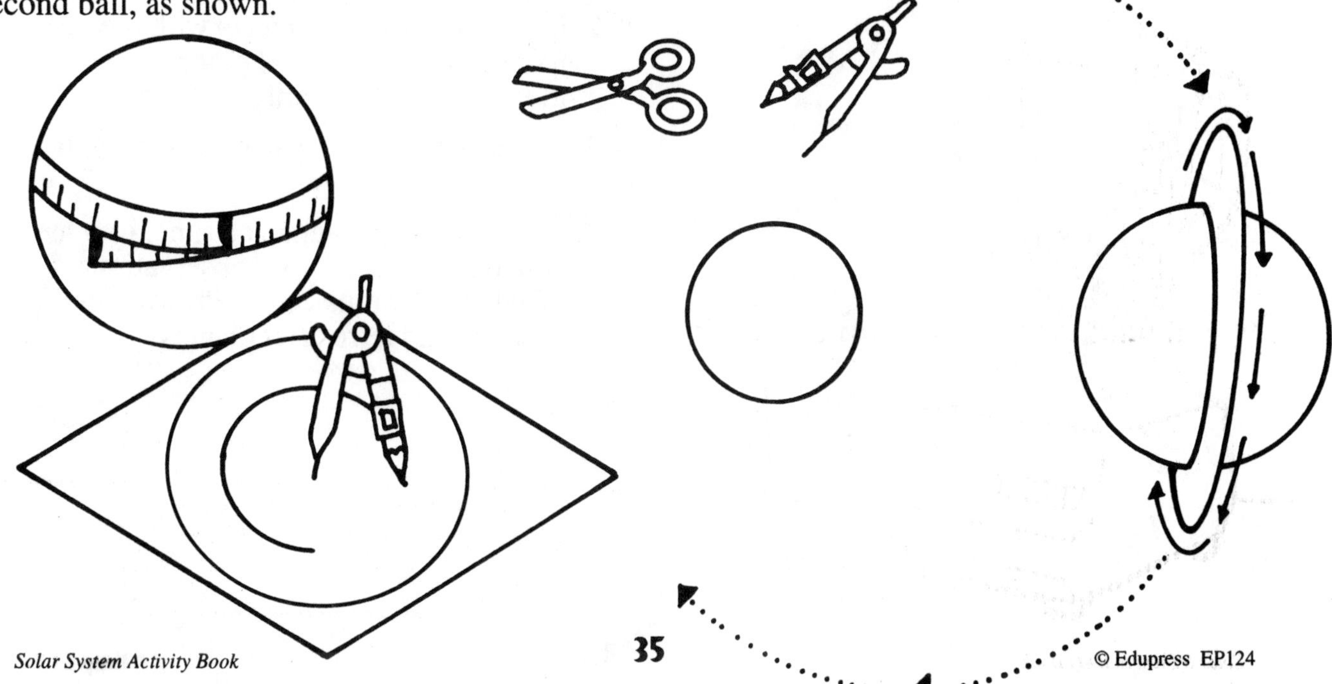

NEPTUNE

INFORMATION

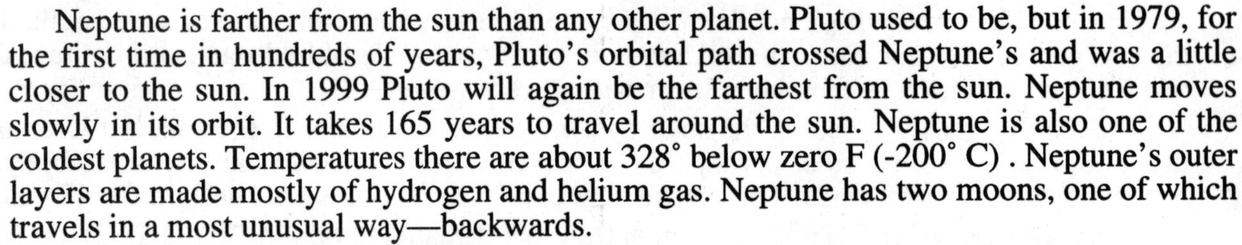

Neptune is farther from the sun than any other planet. Pluto used to be, but in 1979, for the first time in hundreds of years, Pluto's orbital path crossed Neptune's and was a little closer to the sun. In 1999 Pluto will again be the farthest from the sun. Neptune moves slowly in its orbit. It takes 165 years to travel around the sun. Neptune is also one of the coldest planets. Temperatures there are about 328° below zero F (-200° C) . Neptune's outer layers are made mostly of hydrogen and helium gas. Neptune has two moons, one of which travels in a most unusual way—backwards.

The size, or mass, of a planet does not necessarily make it the heaviest. The terrestrial planets have as much as five times the density as the gaseous planets. Neptune is a gaseous planet. One of the largest planets, Saturn, is also the one with the least density!

PROJECT

Conduct a sinking/floating experiment to understand the concept of density.

MATERIALS

- Clear plastic container
- Water
- Permanent marker
- Five floating objects such as ping-pong balls, corks, sponges
- Four sinking objects such as marbles, coins

DIRECTIONS

1. Mark five floating objects with the names of the gaseous planets—Jupiter, Saturn, Uranus, Pluto, and Neptune.
2. Mark four sinking objects with the names of the terrestrial planets—Mercury, Venus, Earth, and Mars.
3. Partially fill the plastic container with water.
4. Drop the planets, one at a time, into the container. Record your observations. What conclusions can be drawn about gaseous and terrestrial planets?

PLUTO

INFORMATION

Pluto is the outermost planet in the solar system. At one point in its orbit it is over 4.5 billion miles (7.24 billion km) from the sun. Pluto is so small and distant from Earth we cannot get a good look at it. It gets hardly any heat and light from the sun. Astronomers believe that Pluto is a freezing and dark world of ice and rock, partly covered with frozen methane gas and little or no atmosphere. Pluto has one moon called Charon.

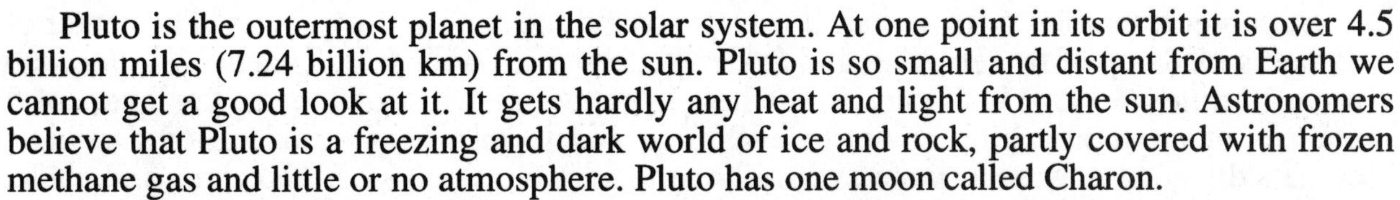

Astronomers are able to tell what the surface of a planet or moon is like by studying sunlight reflecting on its surface. Because Pluto is so far from the sun, the planet and its moon remain worlds of mystery. Astronomers wonder if Pluto is a true planet. Nowhere else in the solar system does the orbit of one planet cross that of another except for Pluto's. Its small size is unusual, too.

PROJECT

Demonstrate the reflective qualities of different materials.

MATERIALS

- Flashlight
- Charcoal briquet
- Cellophane
- Egg
- Orange or lemon
- Aluminum foil
- Rock
- Tissue paper
- Plastic egg or ball

DIRECTIONS

1. Crumple the paper and aluminum foil into balls. Set all the objects a few inches apart in a single line.
2. Darken the room. Shine the flashlight on each object. Observe the reflective characteristics of the materials.
3. Move farther back from the line of objects and repeat the experiment. What differences were observed?
4. Relate your discoveries to the solar system.

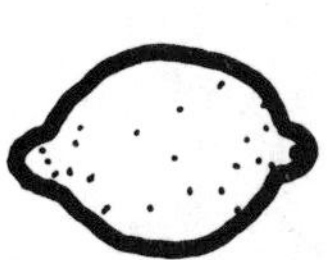

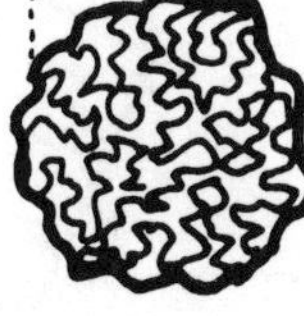

COMET

INFORMATION

A comet resembles a fuzzy star. It travels around the sun in an oval path. The center of a comet is made up of frozen gases and frozen water mixed with dust. This *nucleus* measures up to 10 miles (16 km) in diameter. The heat from the sun causes the outer icy layers of the nucleus of an approaching comet to evaporate and form a hazy cloud called a *coma*. A coma's diameter may be as large as 1 million miles (1.6 million km). The coma and nucleus make up the comet's head. The pressure of the sun's light against the coma forms one or more tails that stream across space as far as 100 million miles (160 million km). When a comet approaches the sun its tail is last, but when it moves away from the sun its tail leads.

Most comets cannot be seen without a telescope. Some are visible to the unaided eye but only for a few weeks or months when they pass closest to the sun.

PROJECT

Make a paper comet. Work in pairs on a comet fact-finding challenge.

MATERIALS

- Paper plate
- Scissors
- Glue
- Comet Challenge page, following
- Resource books

DIRECTIONS

1. Cut a 12-inch (30-cm) length of crepe paper. Fringe it to within two inches (5 cm) of the end. Glue the unfringed end to the paper plate.
2. Take the comets outside and race them in an elliptical orbit around the sun. Remember that as the comet moves away from the sun, its tail leads. You may need a partner to help you achieve this!
3. Reproduce the Comet Fact page. Assign partners for a fact-finding challenge.

COMET CHALLENGE

The time it takes a comet to make a complete orbit around the sun is called its *period*. Some comets have short periods of less than seven years. Others pass near the sun only once in thousands or even millions of years. A comet is constantly in orbit, but we might never see it. There are, however, a few exceptions.

Work with a partner to dig up three interesting facts about comets that have been seen during the period of their orbit. Write the facts below. Share your facts with classmates to learn even more!

HALLEY'S COMET

1.

2.

3.

DONATI'S COMET

1.

2.

3.

COMET KOHOUTEK

1.

2.

3.

ASTEROID

INFORMATION

Asteroids are solar fragments that revolve around the sun between the orbits of Mars and Jupiter. They are constantly tumbling, bumping, and colliding into each other. They come in many sizes and irregular shapes. Asteroids are relatively small—the largest is only six hundred miles (965 km) across.

Asteroids are composed of nickel and iron, or stone, or a little of both. Some may also contain large amounts of carbon which gives them a dull, blackish color. They have little mass compared to Earth. On a very small asteroid you would weigh so little that you might jump up and never come down.

There are thousands of known asteroids. When asteroids collide, as they so often do, they break into pieces. These smaller chunks are called meteoroids.

PROJECT

Create an exhibit of clay asteroids.

MATERIALS

- Self-hardening clay
- Pencils, buttons, toothpicks
- Waxed paper
- Black tempera paint
- Sponge pieces

DIRECTIONS

1. Spread a sheet of waxed paper on the table top.
2. Mold a chunk of clay into any shape. Use pencil erasers, buttons, toothpicks, and other objects to make indentations in the clay. Your fingertip works well, too!
3. Allow the clay to harden.
4. Dip a sponge in black tempera paint and blot the clay.
5. Create an asteroid exhibit with classmates. Are any two asteroids the same in size or shape?

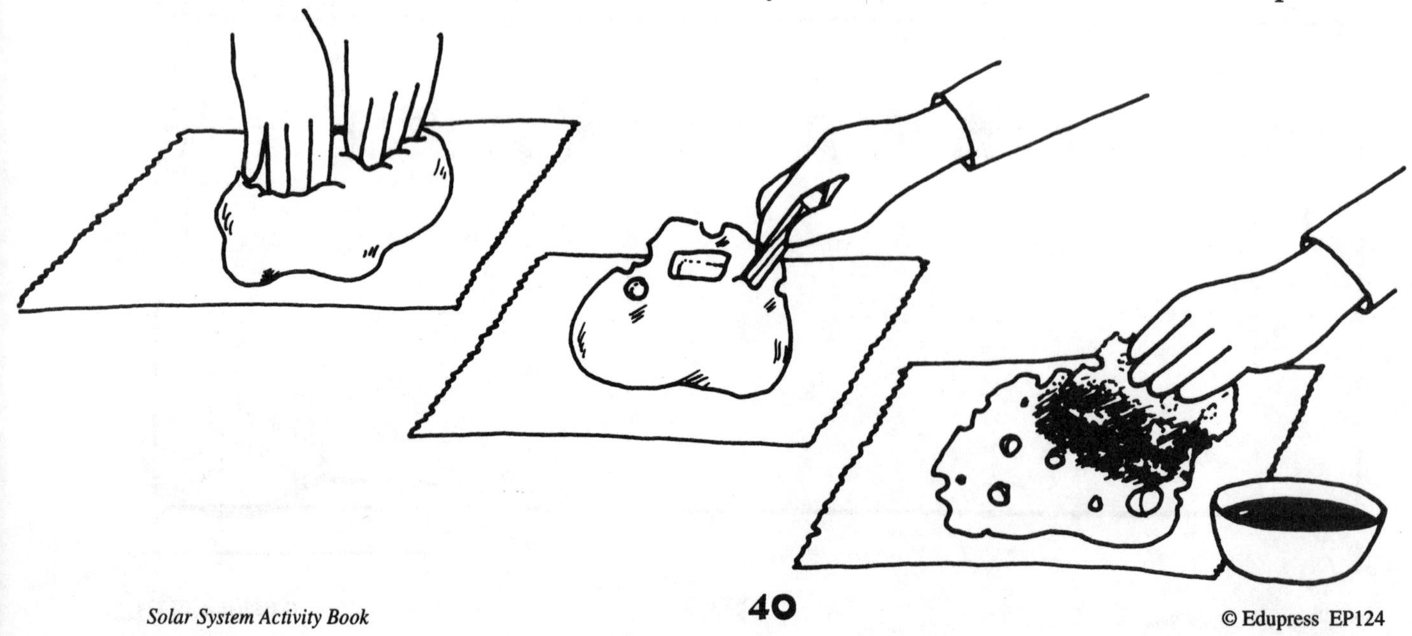

METEOROIDS

INFORMATION

Meteoroids and asteroids are rock or metal chunks that travel at high speed along the orbital paths of comets. When a meteoroid enters the Earth's upper atmosphere, it burns up and forms a trail of hot glowing gases seen as a brief but bright streak of light called a meteor. We also refer to a meteor as a shooting star. As many as 200 million meteors may occur in the Earth's atmosphere every day. Showers of meteors, resulting from the passage of the Earth through the orbit of a comet, occur regularly at certain times of the year.

It is rare that a meteor will strike the Earth's surface. But the moon has no atmosphere to protect it. Meteoroids strike its surface at full speed—20,000 to 160,000 miles (32,000 to 257,000 km) per hour—and make indentations called craters. Some lunar meteoroid explosions have been powerful enough to blast out huge craters nearly one hundred miles (160 km) across.

PROJECT

Experiment with objects to make lunar craters.

MATERIALS

- Flour
- Objects of varying sizes including: marbles, rocks, baseball
- Black butcher paper
- Measuring tape
- Shallow pan or pie tin

DIRECTIONS

1. Fill the pie tin or shallow pan with flour to a one-inch (2.54 cm) level. Set it on the butcher paper.
2. Stand over the pan and choose an object to drop from waist height. Gently lift out the object to examine the crater.
3. Compare the size and weight of each object dropped. Measure the distance of the crater it creates. Measure the distance the flour is scattered onto the butcher paper.
4. Clean off the black butcher paper after every explosion, but leave the crater intact.
5. Examine the surface of the flour at the conclusion of the experiment. Did any of the craters overlap? Imagine the surface of the moon with its constant shower of exploding meteors. Be thankful the Earth's atmosphere is responsible for burning up all the meteors headed our way!

ASTRONOMER

INFORMATION

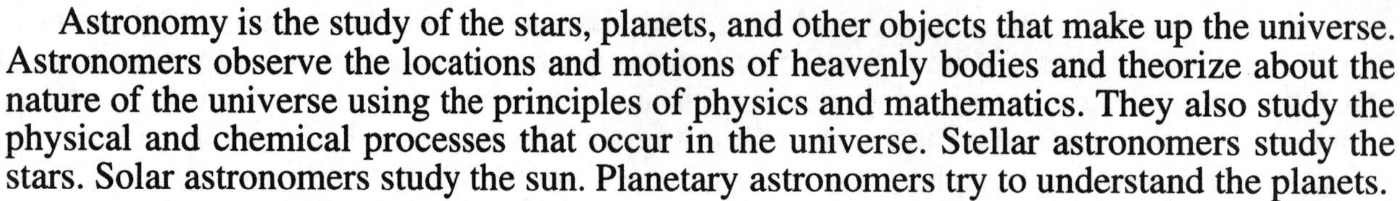

Astronomy is the study of the stars, planets, and other objects that make up the universe. Astronomers observe the locations and motions of heavenly bodies and theorize about the nature of the universe using the principles of physics and mathematics. They also study the physical and chemical processes that occur in the universe. Stellar astronomers study the stars. Solar astronomers study the sun. Planetary astronomers try to understand the planets.

Astronomy is one of the oldest sciences. It began in ancient times. The study of the heavenly cycles has served such practical purposes as keeping time, marking the arrival of the seasons, and navigating the seas.

Astronomers use different types of telescopes to observe many solar system phenomena. They observe light, record radiation patterns, and collect radio waves.

PROJECT

Simulate the role of an astronomer. Hold a news conference to reveal an exciting discovery you've made about the solar system.

MATERIALS

- Astronomer Discovery Cards, following
- Materials as needed by each group

DIRECTIONS

1. Discuss the different types of astronomers.
2. Divide into cooperative groups of three or four.
3. Give each group a Discovery Card.
4. Allow time for each group to:
   - Review the Discovery Card
   - Research pertinent information
   - Imagine a discovery
   - Assemble their materials
   - Create and practice a presentation
5. Each group of astronomers is responsible for holding a news conference about an exciting discovery in the solar system. They should relate how they information was obtained (observation, experimentation, etc.).

ASTRONOMER DISCOVERY CARDS

STELLAR ASTRONOMER

▼ You've discovered a way to bottle star light and create unusual new products.

▼ A new star has been discovered and it's closer to Earth than any other. Are there dangers? benefits?

▼ One star in a constellation has disappeared. What has happened to it?

▼ A new constellation has been identified. Describe its form and location.

PLANETARY ASTRONOMER

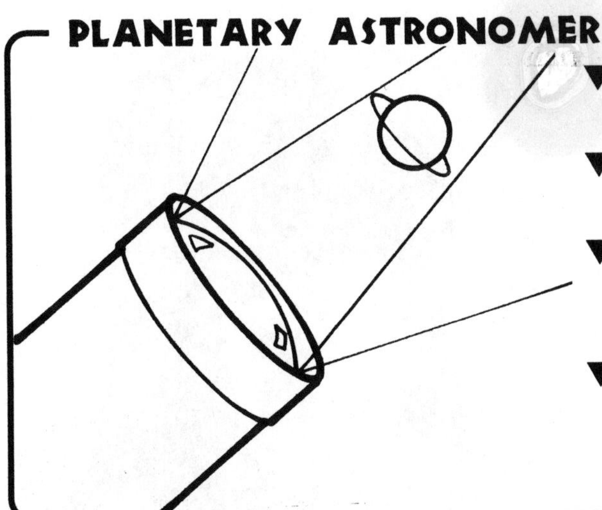

▼ You've discovered signs of life on a distant planet.

▼ A huge comet is headed toward Mars. What can we expect will happen?

▼ A new planet has been located in a high-powered telescope. What information about it can be shared?

▼ Amazing new features have been discovered on the surface of Mars!

SOLAR ASTRONOMER

▼ New spots have appeared on the surface of the sun. What are they?

▼ Something strange is happening to the core of the sun. What observations have been recorded?

▼ A new way has been discovered to measure the sun's temperature.

▼ The size of the sun is changing! What will the effects be here on Earth?

MASTERING THE UNIVERSE

Become masters of the universe with these activities designed to make the language of the solar system as recognizable as everyday words.

KITCHEN CUPBOARD SOLAR SYSTEM

PROJECT

Make a collage of package labels and products named for solar system members.

MATERIALS

- White construction paper
- Magazines
- Scissors
- Glue

DIRECTIONS

1. Look through magazines for pictures of products named for solar system objects. Cut them out.
2. Also look through kitchen cabinets for solar system names. Peel off labels or cut the names from packages. A trip to the market can yield a list of products and drawings.
3. Arrange all the labels, pictures, and product drawings in a collage on construction paper.
4. Glue in place and display.

MASTERING THE UNIVERSE

WORD OF THE DAY

PROJECT

Incorporate solar system words into your daily vocabulary.

MATERIALS

- Butcher paper
- Tape
- Index cards
- Glossary
- Marking pen
- Construction paper star
- Star stickers

DIRECTIONS

1. Tape a large sheet of butcher paper on the wall. Print the glossary words on index cards to create one set. Tape the index cards to the butcher paper.
2. Cut one star from construction paper. Tape the star next to one of the glossary words. Change the position of the star each day.
3. Challenge students to use the word of the day in their conversation. "I was moving faster than a meteor." Award them with a star sticker for each use.

WORD OF THE DAY

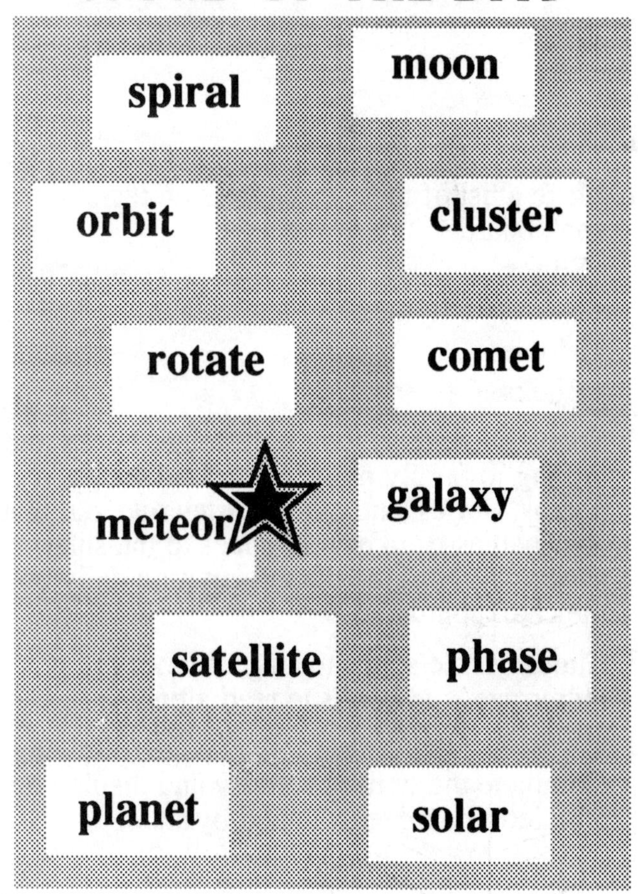

SOLAR WRITING

INFORMATION

The solar system is an inspiration to all authors. We can recognize reference to parts of the solar system in nursery rhymes and song lyrics. Trips to the moon and outer space are the basis for many science fiction plots.

The entertainment industry has also provided us with glimpses into outer space. Some of the most popular movies ever made are about trips to space and life in other galaxies.

PROJECT

Compile a categorical list related to the solar system. Complete solar related writing projects.

MATERIALS

- Chalkboard
- Literary Stars writing page, following

DIRECTIONS

1. Make a list of these categories on the chalkboard:
   - Songs
   - Nursery Rhymes
   - Movies
   - Books
2. Spend a few days looking through books and talking to family members to find songs, poems, stories, and other written material that uses names from or references to the solar system. Record the the information daily on the chalkboard.
3. Choose some songs to sing, rhymes to memorize, and books to read aloud.
4. Reproduce the Literary Stars page following. Complete the writing activity and display on a bulletin board surrounded by stars.

LITERARY STARS

WISH UPON A STAR

Star light, star bright
First star I see tonight
I wish I may, I wish I might
Have this wish come true tonight.

If you could wish upon a star,
what would you wish, and why?

STARRY POETRY

Twinkle, twinkle little star.
How I wonder what you are.
Up above the world so high,
Like a diamond in the sky.

Write a poem about an object in the solar system. Here are some ideas to get you started

- ☆ Far Away Mars
- ☆ Fun in the Sun
- ☆ Spin, Planet, Spin
- ☆ Marvelous Moonshine

WORLD WIDE WEB

Look in the world wide web to expand your knowledge of the solar system. Keep in mind that web pages change constantly. The web pages below were active at publication date but their continued presence is not guaranteed. Like the universe, however, the world wide web is constantly expanding, just waiting for budding astronomers to make exciting discoveries.

ADDRESS	CONTENT
itss.raytheon.com/cafe/qadir/qanda.html	Ask the Astronomer—check the archives of 3,001 questions already answered or submit your own. This site is part of the Astronomy cafe.
www.theastronomycafe.net	Astronomy Cafe—astronomy career guide, big bang cosmology, other Web resources and "the infrared universe" (technical details of astrophysics).
webhome.idirect.com/~rsnow/	Absolute Beginners Astronomy and Telescopes—a question-and-answer section on buying your first telescope and how to locate objects in the sky.
www.seds.org/nineplanets/nineplanets/nineplanets.html	The Nine Planets—an overview of the history, mythology, and current scientific knowledge of each of the planets and moons in our solar system.
www.ias.fr/cdp/solar/eng/homepage.htm	View of the Solar System—multimedia adventure unfolding the splendor of the sun, planets, moons, comets, asteroids, and more.

